Earth

The Living Paradox

JD ARDEN

Preface: A Singular World

Earth is a marvel. Of the countless worlds scattered across the cosmos, it is the only one we know of that teems with life. From the deepest oceans to the highest peaks, life flourishes in countless forms, adapting to every corner of this dynamic planet. Earth's uniqueness is not just in its ability to host life but in the complexity of the systems that sustain it—a balance so intricate that it feels both inevitable and miraculous.

Earth sits at the perfect distance from the Sun, cradled in the habitable zone where temperatures allow liquid water to persist. It is shielded by a magnetic field and an atmosphere that protects it from cosmic radiation and solar winds. Its oceans moderate the climate, its tectonic plates drive evolution and renew its surface, and its Moon stabilizes its axis, ensuring seasonal cycles that nurture biodiversity. Together, these factors create a delicate equilibrium, a harmony that has allowed life to flourish for billions of years.

Yet, Earth is also a paradox. Its stability masks a history of upheaval—asteroid impacts, mass extinctions, and ice ages that have reshaped the planet again and again. Life itself, while thriving, has altered the planet irrevocably, from the oxygenation of the atmosphere to the vast human footprint of the Anthropocene. Earth is a world of contrasts: fragile yet enduring, predictable yet dynamic, a cradle for life and a force of transformation.

This book explores Earth not just as a home but as a singular phenomenon, a living paradox in the vastness of space. It is a celebration of the systems that sustain life, an inquiry into the forces that shape our world, and a reflection on the choices we face as stewards of this fragile planet.

Chapter 1: A Perfect Distance

Earth's position in the solar system is often described as "just right." Not too close to the Sun, where searing heat would strip away its atmosphere and boil away its water, nor too far, where icy desolation would make liquid water impossible. Earth resides in the **habitable zone**, sometimes called the Goldilocks zone—a narrow band where conditions are favorable for life as we know it. This positioning is the foundation upon which all other life-sustaining systems are built.

The concept of the habitable zone is deceptively simple. It is defined by the range of distances from a star where temperatures allow liquid water to exist. For Earth, this zone extends roughly from just inside Venus's orbit to just beyond Mars's. But the reality is far more nuanced. A planet's habitability depends not just on its distance from its star but also on factors like atmospheric composition, geological activity, and the characteristics of the star itself. Earth, it seems, occupies a perfect confluence of conditions.

The Sun, for instance, is a stable star of medium size and luminosity. It emits energy at a consistent rate, avoiding the violent outbursts and variability that characterize smaller red dwarfs or massive blue stars. This consistency has allowed Earth's climate to remain relatively stable over geological timescales, providing a steady environment for life to evolve.

Earth's orbit is another stroke of luck. It is nearly circular, ensuring that the planet stays within the habitable zone year-round. Planets with more elliptical orbits experience extreme temperature swings, moving in and out of their stars' habitable zones and creating environments that are far less conducive to life.

Earth's axial tilt, stabilized by the gravitational influence of the Moon, is equally critical. This tilt, at an angle of about 23.5 degrees, creates the seasonal variations that moderate the planet's climate. Without this tilt, Earth's poles might freeze solid while the equator baked, or its climate could oscillate wildly over time. The Moon's role in stabilizing this tilt ensures that seasons change predictably, providing a consistent rhythm for ecosystems to adapt and thrive.

Earth

But Earth's position is more than just a matter of geography—it is a balance of energy. The planet receives energy from the Sun, which warms its surface and drives weather patterns, but it also radiates energy back into space. This energy balance is finely tuned. The atmosphere, with its mix of greenhouse gases like carbon dioxide and water vapor, traps just enough heat to keep the planet warm without overheating. A slight increase or decrease in this greenhouse effect could tip Earth into a frozen ice age or a sweltering greenhouse state, as has happened during various points in its history.

Water, often called the lifeblood of Earth, is perhaps the most visible expression of this balance. Liquid water exists here because of the planet's position in the habitable zone, but also because of the atmospheric and geological systems that regulate its presence. Water absorbs heat, moderates temperature, and drives the weather cycle that redistributes heat and moisture across the planet. It is no exaggeration to say that life on Earth depends entirely on the availability of liquid water, made possible by its location and climate.

While Earth's position seems perfectly suited for life, it is worth remembering that the habitable zone is not a static concept. The Sun is slowly growing brighter as it ages, expanding the habitable zone outward. Over billions of years, this gradual brightening will push Earth closer to the inner edge of the zone, making it too hot for liquid water to persist. Venus, which may have once had oceans and a temperate climate, is a stark reminder of what happens when a planet slips out of this delicate balance.

Earth's perfect distance from the Sun also raises questions about the rarity of life elsewhere in the universe. Astronomers have identified thousands of exoplanets, many of them in their stars' habitable zones. But habitability is more than a matter of distance; it is the result of countless factors working in harmony. Earth's example reminds us that while habitable zones may be common, the precise combination of conditions that sustain life is extraordinarily rare.

The Goldilocks principle—that Earth is "just right"—is both a scientific observation and a philosophical reflection. It highlights the interconnectedness of the systems that make life possible and the fragility of the balance that sustains them. Earth's position is not a

Earth

guarantee of habitability; it is an opportunity, one that has been shaped and maintained by billions of years of cosmic, geological, and biological processes.

In its position, Earth embodies the paradox of stability and change. It orbits in a predictable path, bathed in sunlight and shielded by a magnetic field, yet it is a dynamic world, constantly evolving, its surface reshaped by tectonics, erosion, and life itself. This balance—of constancy and transformation, of proximity and distance—is what makes Earth a living paradox, a world unlike any other we have yet encountered.

Earth's place in the habitable zone is not just a scientific fact; it is a reminder of the preciousness of our planet. It is a testament to the intricate, delicate systems that allow life to exist, and a call to appreciate and protect the unique world we call home.

Chapter 2: The Atmosphere's Blanket

Earth's atmosphere is one of its most defining features, a delicate yet robust system that has evolved over billions of years to shield, nurture, and sustain life. From protecting the surface from harmful solar radiation to regulating temperature and enabling the water cycle, the atmosphere is the unsung hero of Earth's habitability. Without it, our planet would be an inhospitable wasteland, resembling the barren deserts of Mars or the searing hellscape of Venus.

To understand Earth's atmosphere is to appreciate the intricate interplay of gases, energy, and processes that have transformed a once-hostile environment into a cradle for life. It is not merely a layer of air; it is a dynamic system, constantly reshaped by geological forces, biological activity, and even human influence.

The story of Earth's atmosphere begins with its formation around 4.6 billion years ago. In its earliest days, Earth's atmosphere was vastly different from the one we know today. This **primordial atmosphere**, formed from gases released during the planet's accretion and early volcanic activity, was dominated by hydrogen and helium, the lightest and most abundant elements in the universe. But this atmosphere was transient. The Sun's intense radiation and solar wind quickly stripped these light gases away, leaving Earth bare and exposed.

The second atmosphere emerged as the planet cooled and volcanic eruptions spewed gases from Earth's molten interior. This atmosphere was rich in carbon dioxide, methane, ammonia, nitrogen, and water vapor, creating a thick, greenhouse-like environment. There was virtually no oxygen, and the planet's surface was dominated by molten rock and cooling oceans. While this atmosphere was inhospitable to life as we know it, it set the stage for the dramatic transformations to come.

A pivotal moment in Earth's atmospheric history occurred around 3.5 billion years ago with the emergence of life. Cyanobacteria, the earliest photosynthetic organisms, began to harness sunlight to convert carbon dioxide and water into energy, releasing oxygen as a byproduct. This biological process initiated the **Great Oxygenation Event** (GOE) around 2.4 billion years ago, a watershed moment that would redefine Earth's atmosphere and its habitability.

As oxygen accumulated, it reacted with methane in the atmosphere, replacing a potent greenhouse gas with one less efficient at trapping heat. This triggered a series of global glaciations, known as "Snowball Earth" events, as temperatures plummeted. Yet, the oxygen also paved the way for the formation of the **ozone layer**, a shield of O_3 molecules that absorbed harmful ultraviolet radiation from the Sun. The ozone layer was critical for life's transition from the oceans to land, protecting organisms from the Sun's deadly rays and enabling the diversification of complex life.

Over time, Earth's atmosphere stabilized into its current composition: approximately 78% nitrogen, 21% oxygen, and trace amounts of other gases, including argon, carbon dioxide, and water vapor. Each of these components plays a crucial role. Nitrogen provides stability, diluting oxygen to prevent widespread combustion, while oxygen fuels the metabolism of most living organisms. Trace gases, though present in tiny amounts, have outsized effects. Carbon dioxide and water vapor, for example, are essential for maintaining the greenhouse effect, which keeps Earth warm enough to support life.

The atmosphere's structure is equally remarkable, divided into layers, each with distinct properties and functions. The **troposphere**, the lowest layer, is where weather occurs and where most life exists. Rising to an altitude of about 12 kilometers, it is characterized by a decrease in temperature with height and contains the bulk of the atmosphere's water vapor. The troposphere's dynamic weather patterns—driven by the Sun's energy—redistribute heat and moisture across the planet, sustaining ecosystems and moderating the climate.

Above the troposphere lies the **stratosphere**, home to the ozone layer. Unlike the troposphere, the stratosphere experiences an increase in temperature with altitude, as ozone absorbs ultraviolet radiation and converts it into heat. This layer acts as a protective shield, filtering out the majority of the Sun's harmful UV rays and stabilizing the atmosphere below.

Beyond the stratosphere, the **mesosphere** and **thermosphere** play critical roles in shielding Earth from meteoroids and solar radiation. The mesosphere, where temperatures plunge to their lowest, causes most meteoroids to burn up upon entry, creating the streaks of light we see as shooting stars. The thermosphere, meanwhile, absorbs high-energy X-rays and ultraviolet radiation, causing temperatures to soar and creating the auroras seen near the poles.

At the edge of space lies the **exosphere**, a thin layer where atmospheric particles drift into the vacuum of space. While tenuous, this layer marks the transition between Earth's atmosphere and the interplanetary medium, serving as a boundary for Earth's protective cocoon.

The atmosphere's role in regulating Earth's climate cannot be overstated. Through the greenhouse effect, it traps just enough of the Sun's energy to maintain a stable, life-supporting temperature. However, this balance is delicate. An increase in greenhouse gases, whether through natural processes like volcanic eruptions or human activities like burning fossil fuels, can disrupt this equilibrium, leading to global warming. Venus, with its thick, CO_2-dominated atmosphere, offers a cautionary tale of what happens when the greenhouse effect spirals out of control.

In addition to regulating temperature, the atmosphere drives the **water cycle**, a system that redistributes heat and moisture around the globe. Evaporation, condensation, and precipitation are powered by the Sun's energy, linking the atmosphere to Earth's oceans, rivers, and ice caps. This cycle is essential for sustaining ecosystems, shaping weather patterns, and replenishing freshwater supplies.

Earth's atmosphere is not static; it is a dynamic system that has changed dramatically over geological time and continues to evolve today. Human activities are now one of the most significant drivers of atmospheric change, from the buildup of greenhouse gases to the depletion of the ozone layer. These changes, though often subtle on a day-to-day scale, have profound implications for the planet's future.

Despite its fragility, Earth's atmosphere is a testament to the resilience and adaptability of life. It is both a product of the planet's geological and biological processes and a protector of the ecosystems that depend on it.

Earth

It connects the deep past to the present, offering clues about the conditions that allowed life to emerge and thrive.

The atmosphere is more than a layer of gases; it is a living system, a guardian, and a record of Earth's history. As we look to other planets, the study of atmospheres—both ours and theirs—deepens our understanding of what makes Earth so unique. It reminds us that the conditions for life are not guaranteed but are the result of billions of years of intricate, interwoven processes.

In its balance of protection and dynamism, Earth's atmosphere exemplifies the paradox of stability and change, offering a living testament to the planet's singular ability to sustain life.

Chapter 3: Oceans of Possibility

Earth's oceans are its lifeblood, covering more than 70% of the planet's surface and forming a vast, interconnected system that supports life, regulates climate, and shapes the planet's very identity. Without the oceans, Earth would not be the vibrant, dynamic world we know today. They are more than reservoirs of water; they are engines of life and stability, repositories of memory and change.

The origins of Earth's oceans are rooted in the planet's earliest history. Billions of years ago, as Earth cooled from its molten beginnings, water vapor released by volcanic activity began to condense, forming the first pools and seas. Additional water likely arrived from icy comets and asteroids that bombarded the planet during its formative years, enriching the surface with hydrogen and oxygen compounds. These primordial waters collected into basins, creating the oceans that would come to define Earth's surface.

From the moment they formed, Earth's oceans became agents of transformation. They absorbed heat, regulated temperatures, and provided the medium in which the building blocks of life could assemble. The first lifeforms—microscopic organisms that emerged more than 3.5 billion years ago—evolved in the nutrient-rich waters of Earth's early seas, shielded from the harsh ultraviolet radiation of the Sun by the water above.

The oceans are not just a cradle for life; they are its sustainer. Today, marine ecosystems support an astonishing diversity of organisms, from microscopic plankton to the largest whales. These ecosystems form the foundation of Earth's food web, with phytoplankton—tiny, photosynthetic organisms—serving as the primary producers that convert sunlight into energy. Phytoplankton also play a crucial role in Earth's carbon cycle, absorbing vast amounts of carbon dioxide from the atmosphere and releasing oxygen. It is no exaggeration to say that without the oceans, the atmosphere as we know it would not exist.

Beyond their role as life-sustainers, the oceans are Earth's most powerful climate regulators. Water has an extraordinary capacity to store and transfer heat, making the oceans a global thermostat that moderates temperatures across the planet. Ocean currents, driven by wind, salinity, and temperature gradients, transport warm water from the equator to the poles and cold water back to the tropics. This system, known as the **thermohaline circulation** or the "global conveyor belt," is essential for maintaining Earth's climate stability.

One of the most dramatic examples of the ocean's influence on climate is the **El Niño-Southern Oscillation (ENSO)**. During El Niño events, warmer-than-average waters in the equatorial Pacific disrupt weather patterns worldwide, leading to floods, droughts, and shifts in global temperatures. Conversely, La Niña events, characterized by cooler-than-average waters, have opposite but equally profound effects. These phenomena illustrate how intimately Earth's climate is tied to the behavior of its oceans.

The oceans also act as a buffer against climate change, absorbing about 90% of the excess heat trapped by greenhouse gases and more than a quarter of the carbon dioxide emitted by human activities. While this has slowed the pace of atmospheric warming, it comes at a cost. The absorption of CO_2 leads to **ocean acidification**, a process that reduces the pH of seawater and threatens marine ecosystems, particularly coral reefs and shell-forming organisms. Similarly, the heat absorbed by the oceans contributes to thermal expansion and melting ice, driving **sea level rise** that threatens coastal communities worldwide.

Despite their vastness, Earth's oceans are deeply interconnected, forming a single, continuous body of water that flows around the globe. This interconnectedness is a source of strength and vulnerability. Pollution introduced in one region, such as plastic waste or oil spills, can spread across entire ocean basins, affecting ecosystems thousands of kilometers away. Overfishing, habitat destruction, and climate change further compound these challenges, highlighting the need for global stewardship of the oceans.

The geological impact of the oceans is equally profound. Over millions of years, the movement of tectonic plates beneath the ocean floor has created vast mid-ocean ridges, subduction zones, and deep-sea trenches. These underwater features drive the cycling of nutrients and minerals, fueling life in even the darkest depths. Hydrothermal vents, for example, host unique ecosystems powered by chemical energy rather than sunlight, proving that life can thrive in extreme conditions.

Earth's oceans are also a repository of memory, preserving clues about the planet's past in their sediments and chemical composition. By studying ice cores, deep-sea sediments, and coral reefs, scientists can reconstruct ancient climates, uncovering patterns of change that span millions of years. These records reveal the resilience of Earth's systems but also their vulnerability to sudden shifts.

The philosophical significance of Earth's oceans is as profound as their physical impact. They are a symbol of interconnectedness, a reminder that life is sustained by the intricate web of relationships between water, air, land, and organisms. They challenge us to think beyond borders and consider the planet as a whole, united by the currents and tides that flow ceaselessly across its surface.

The oceans also inspire awe and humility. Their vastness, depth, and mystery remind us of how much remains unexplored. Despite centuries of study, more than 80% of the ocean floor remains unmapped, and new species are discovered regularly in its depths. The ocean is a frontier, a realm of endless discovery that mirrors the boundless creativity of nature.

Yet, the oceans are not invulnerable. Their health is directly tied to the actions of humanity, from the emissions we produce to the resources we extract. Climate change, pollution, and overexploitation are altering the oceans in ways that could have cascading effects on the entire planet. Protecting the oceans is not just an environmental imperative; it is a matter of survival for life on Earth.

Earth's oceans are a paradox of strength and fragility, constancy and change. They have endured for billions of years, shaping the planet and nurturing life, yet they are acutely sensitive to disruption. They remind us of the delicate balance that sustains our world and the responsibility we bear as stewards of this interconnected system.

In their depths, currents, and lifeforms, the oceans tell the story of Earth itself—a story of origins, resilience, and the ongoing dance between

Earth

stability and transformation. They are a testament to the creativity of the cosmos and the singularity of our living planet.

Chapter 4: The Dance of Plates

Beneath the rolling oceans and the forests, plains, and mountains of Earth's surface lies a restless, dynamic system that shapes the planet in ways both subtle and dramatic. This system, known as **plate tectonics**, is unique among the planets in our solar system—a phenomenon that not only sculpts the physical features of Earth but also plays a pivotal role in its habitability. Plate tectonics is Earth's engine of renewal, a driving force behind earthquakes, volcanoes, and the formation of continents. It is also a critical component of the processes that sustain life.

The concept of plate tectonics is relatively recent in the history of science, emerging as a unifying theory in the mid-20th century. It describes how Earth's outer shell, the **lithosphere**, is broken into massive slabs, or plates, that float atop the semi-fluid **asthenosphere**, a layer of partially molten rock in the upper mantle. These plates—seven major ones and numerous smaller ones—are in constant motion, propelled by the convective currents of heat rising from Earth's interior.

The movement of tectonic plates drives a wide range of geological processes, from the formation of mountain ranges to the opening and closing of ocean basins. At the boundaries where plates interact, their motion creates some of Earth's most dramatic phenomena. **Divergent boundaries**, where plates move apart, give rise to mid-ocean ridges like the Atlantic's Mid-Atlantic Ridge, where new crust is formed as magma rises to the surface. **Convergent boundaries**, where plates collide, produce mountain ranges such as the Himalayas, formed by the ongoing collision of the Indian and Eurasian plates. In some cases, one plate is forced beneath another in a process called **subduction**, creating deep ocean trenches like the Mariana Trench and fueling explosive volcanic activity along subduction zones.

Transform boundaries, where plates slide past one another, generate the intense friction that causes earthquakes. The San Andreas Fault in California is one of the most famous examples of this kind of boundary, where the Pacific and North American plates grind past each other in fits and starts, unleashing seismic energy that ripples through the landscape.

Earth

While these processes often result in dramatic and destructive events, they are also essential for the planet's long-term stability. Plate tectonics plays a critical role in the **carbon cycle**, a system that regulates the levels of carbon dioxide in Earth's atmosphere and keeps global temperatures in check. When tectonic plates subduct, they carry carbon-rich sediments into the mantle, where some of the carbon is sequestered and some is released back into the atmosphere through volcanic eruptions. This cycle acts as a planetary thermostat, helping to stabilize Earth's climate over millions of years.

Plate tectonics also drives the recycling of Earth's crust, renewing the surface and preventing the planet from becoming geologically stagnant. Without this renewal, Earth's surface might resemble the Moon or Mars—scarred by ancient craters and devoid of the dynamic processes that support ecosystems. By creating new land and reshaping existing landscapes, plate tectonics fosters biodiversity, offering varied environments for life to adapt and thrive.

The relationship between plate tectonics and life is profound. The shifting of plates and the creation of new landmasses have driven the evolution and distribution of species throughout Earth's history. When continents drift apart, species are isolated, leading to speciation and the development of unique ecosystems. When continents collide, the merging of habitats can create new opportunities for life to flourish. This interplay between geology and biology has been a driving force behind the diversity of life on Earth.

Plate tectonics also influences the distribution of resources that are essential for human civilization. Many of the planet's richest deposits of minerals, metals, and fossil fuels are the result of tectonic processes. Volcanic activity creates deposits of precious metals like gold and silver, while subduction zones are often associated with oil and gas reserves. Even the soil that supports agriculture owes its fertility to the processes of weathering and nutrient cycling, which are tied to tectonic activity.

But plate tectonics is not without its dangers. The very processes that renew the planet and sustain life also bring destruction. Earthquakes, volcanic eruptions, and tsunamis—while natural and essential—pose significant risks to human populations, particularly in regions located along tectonic boundaries. Understanding these processes and their patterns is critical for mitigating their impacts and protecting communities.

Earth's plate tectonics is not only a unique feature of our planet but also a rare phenomenon in the cosmos. While other planets and moons exhibit tectonic-like activity, such as volcanic resurfacing on Venus or cryovolcanism on Jupiter's moon Europa, none display the global, interconnected system seen on Earth. The absence of plate tectonics on other worlds underscores its importance in creating and sustaining the conditions necessary for life.

The driving force behind plate tectonics lies deep within Earth's interior, where heat generated by radioactive decay and the residual energy from the planet's formation creates convection currents in the mantle. These currents act like a conveyor belt, moving the plates across the surface over geological timescales. This movement is imperceptible in human terms—most plates move at a rate of a few centimeters per year, about the speed at which fingernails grow—but its cumulative effects are monumental.

Philosophically, plate tectonics offers a powerful metaphor for change and renewal. It demonstrates how dynamic processes can create both destruction and opportunity, reshaping the world in ways that foster growth and adaptation. The constant movement of the plates, their collisions and separations, remind us that even seemingly stable systems are in flux, and that transformation is a fundamental aspect of existence.

The dance of plates is a story of contrasts: creation and destruction, stability and upheaval, predictability and surprise. It is a testament to the interconnectedness of Earth's systems, showing how deep geological processes shape the surface world we inhabit. Plate tectonics is not just a geological phenomenon; it is a force that connects the depths of the Earth to the life that flourishes above, weaving the planet's past, present, and future into a single, dynamic narrative.

As we study plate tectonics, we gain not only a deeper understanding of Earth but also a greater appreciation for the delicate balance that makes it unique. It is a reminder of the planet's extraordinary capacity for renewal and the forces that continue to shape the world we call home.

Earth

Chapter 5: The Moon's Silent Influence

Above Earth's surface, the Moon orbits in quiet majesty, a constant companion that has shaped the planet in profound and often unseen ways. While its silvery glow inspires poetry and wonder, the Moon's influence extends far beyond its beauty. It has been a stabilizer, a timekeeper, and an architect of life's rhythms. Without the Moon, Earth would be a dramatically different world, its landscapes, climate, and even the trajectory of life itself altered in unimaginable ways.

The Moon formed roughly 4.5 billion years ago, likely as a result of a cataclysmic collision between the young Earth and a Mars-sized body often referred to as **Theia**. This impact ejected a massive amount of debris into space, which coalesced over time into the Moon. The Moon's proximity to Earth in its early days was far closer than it is today—appearing much larger in the sky and exerting even stronger tidal forces on the planet. Over the eons, these forces have shaped Earth's rotation, its axis, and its biological rhythms.

One of the Moon's most critical roles has been the stabilization of Earth's axial tilt. Earth's axis, tilted at approximately 23.5 degrees, is what gives the planet its seasons. Without the Moon, gravitational interactions with the Sun and other planets, particularly Jupiter, could cause the tilt to wobble chaotically over time, fluctuating between extremes. Such instability would lead to severe climatic swings, with poles alternately freezing and thawing, disrupting ecosystems and potentially making Earth far less hospitable to life.

The Moon's steady gravitational pull has acted as a stabilizer, anchoring Earth's tilt and ensuring the predictability of seasons. This consistency has allowed ecosystems to adapt and thrive, from the regular blooming of plants to the migratory patterns of animals that depend on seasonal changes. The Moon's silent influence, then, has been a key factor in creating a stable environment where life could flourish.

The Moon's gravitational pull also gives rise to Earth's **tides**, a phenomenon that has shaped coastlines, driven evolution, and created unique habitats. Tides occur as the Moon's gravity pulls on Earth's oceans, causing the water to bulge toward and away from the Moon. These rhythmic movements of water have been a constant presence for billions of years, influencing marine ecosystems and creating intertidal zones—areas where the land is alternately submerged and exposed.

These intertidal zones are among the most biologically rich environments on Earth. Early in Earth's history, they may have served as transitional habitats, where life first began to explore the challenges and opportunities of living on land. The ebb and flow of tides also played a role in the dispersal of nutrients, supporting the growth of microbial life and later, complex organisms.

The Moon's role as a timekeeper is another aspect of its profound influence. In the early days of Earth's history, when the Moon was much closer, its gravitational forces slowed Earth's rotation, gradually lengthening the length of a day. From the chaotic 6-hour days of the planet's youth, Earth's rotation has stabilized into the 24-hour cycle we know today. This lengthening of the day created more stable periods of light and darkness, which became essential for the evolution of circadian rhythms in plants, animals, and humans.

Even today, the Moon continues to serve as a regulator of time, influencing the behavior of countless species. Many marine organisms, for example, synchronize their reproductive cycles with the lunar phases, using the Moon's light and tidal patterns as cues. For indigenous cultures and early civilizations, the Moon's predictable cycles provided a natural calendar, guiding agricultural practices, rituals, and navigation.

The Moon's influence extends even to Earth's geological history. The tides it generates create friction as they interact with Earth's rotation, gradually transferring energy from the planet to the Moon. This energy transfer causes the Moon to slowly drift away from Earth—at a rate of about 3.8 centimeters per year. Over geological timescales, this separation affects the dynamics of Earth's rotation and its relationship with the Moon, subtly altering the length of a day and the magnitude of tidal forces.

In addition to its gravitational effects, the Moon has played a role in Earth's geological processes through impacts. The craters that pockmark

Earth

the Moon's surface are a reminder of the shared history of bombardment that Earth and the Moon experienced during the early solar system. While Earth's atmosphere and active geology have erased many of its ancient craters, the Moon's surface serves as a fossilized record of those tumultuous times. The debris from these impacts may have even contributed to the delivery of water and organic materials to Earth, seeding the planet with the ingredients for life.

Philosophically, the Moon embodies the paradox of presence and silence. Though it exerts such profound influence on Earth, it does so quietly, without visible force. It reminds us that the most significant forces in the universe are often the least conspicuous. The Moon's consistent orbit and cycles are a source of comfort and continuity, a reminder of Earth's connection to the cosmos and the delicate balances that sustain life.

The Moon's influence also challenges us to consider the interconnectedness of systems on Earth and beyond. It is not merely an isolated celestial body but an integral part of the Earth-Moon system, a partnership that has shaped the trajectory of our planet and the evolution of its inhabitants. This relationship underscores the idea that life on Earth is not just a product of terrestrial forces but of cosmic interactions that extend far beyond the planet's surface.

As humanity looks to the stars, the Moon remains a constant, a companion that has shaped our past and inspires our future. Its surface, marked by ancient impacts and untouched by weather, holds the keys to understanding the history of the solar system. Its gravitational pull continues to influence Earth's rotation, tides, and rhythms, quietly reminding us of the unseen forces that shape our world.

The Moon's silent influence is a testament to the power of connection—a force that binds worlds together and shapes their destinies. It is a partner in Earth's journey, a stabilizer in a chaotic cosmos, and a symbol of the profound relationships that sustain life in the universe.

Chapter 6: The Oxygen Revolution

Earth's atmosphere, with its life-sustaining 21% oxygen, is a cornerstone of the planet's habitability. Yet this abundance of oxygen was not always a given. For much of Earth's early history, the atmosphere was devoid of this vital gas, consisting instead of carbon dioxide, methane, nitrogen, and other compounds. The journey from an oxygen-poor world to the oxygen-rich environment we know today is one of the most transformative events in Earth's history, a revolution that reshaped the planet's atmosphere, geology, and biosphere.

The story of the **Oxygen Revolution** begins with the emergence of life. Around 3.5 billion years ago, microscopic organisms called cyanobacteria began to perform **photosynthesis**, a process that used sunlight to convert carbon dioxide and water into organic compounds, releasing oxygen as a byproduct. This was a slow and subtle process at first, as oxygen readily reacted with the abundant iron and other elements in the oceans, forming insoluble compounds that precipitated to the seafloor. These reactions created the **banded iron formations** found in ancient rock layers, evidence of a time when Earth's oceans were chemically transformed by the appearance of oxygen.

For nearly a billion years, oxygen produced by cyanobacteria was absorbed by these chemical sinks, preventing it from accumulating in the atmosphere. But around 2.4 billion years ago, this balance shifted. The **Great Oxygenation Event (GOE)** marked the point when these sinks became saturated, allowing oxygen to build up in the atmosphere for the first time. This was not a gentle transformation; it was a global upheaval that had profound and often catastrophic consequences for life and the planet.

The sudden rise of atmospheric oxygen, though vital for the development of complex life, was toxic to many of the anaerobic organisms that had dominated Earth up to that point. For these oxygen-intolerant microbes, the GOE was an extinction event, wiping out vast populations and altering the trajectory of evolution. But for other organisms, the presence of oxygen opened up new possibilities. The ability to harness oxygen for **aerobic respiration**, a process far more efficient than anaerobic pathways, allowed life to grow larger, more complex, and more diverse.

Oxygen's influence extended far beyond biology. Its presence in the atmosphere triggered a series of chemical and geological changes that transformed the planet itself. One of the most significant was the formation of the **ozone layer**, a protective shield of O_3 molecules in the stratosphere. The ozone layer absorbs the majority of the Sun's harmful ultraviolet radiation, creating a safer environment for life to venture out of the oceans and onto land. Without this shield, life's expansion onto terrestrial environments would have been far more difficult, if not impossible.

The rise of oxygen also changed Earth's climate. By reacting with methane—a potent greenhouse gas—oxygen reduced the atmosphere's ability to trap heat, leading to a series of global glaciations known as "Snowball Earth" events. These ice ages, though extreme, may have driven evolutionary innovation by creating new environmental pressures and challenges for life to overcome.

Over the next billion years, atmospheric oxygen levels fluctuated, influenced by biological, geological, and climatic factors. The balance between oxygen production through photosynthesis and its consumption through respiration and decay created a dynamic system, one that continues to evolve today. Around 600 million years ago, during the **Neoproterozoic Era**, oxygen levels rose dramatically again, coinciding with the appearance of the first multicellular animals. This surge in oxygen availability is thought to have been a critical factor in enabling the **Cambrian Explosion**, a period of rapid evolutionary diversification that produced the ancestors of most major animal groups.

The Oxygen Revolution also highlights the deep interconnections between Earth's systems. It demonstrates how life and the planet's physical environment are not separate entities but partners in a dynamic, co-evolving relationship. The oxygen produced by cyanobacteria not only altered the atmosphere but also reshaped Earth's oceans, influencing their chemistry, circulation, and capacity to support life. This, in turn, affected the distribution and evolution of marine ecosystems, setting the stage for the incredible diversity of life we see today.

From a philosophical perspective, the Oxygen Revolution challenges us to reconsider the nature of transformation. It was a slow, incremental process that led to dramatic, even catastrophic, consequences—a reminder that profound change often arises from the cumulative effects of small, seemingly inconsequential actions. It also underscores the paradox of progress: oxygen, while essential for complex life, was toxic to the organisms that came before. Progress, in this sense, is not linear but layered, with each step forward building upon the destruction and reinvention of what came before.

The Oxygen Revolution is also a testament to the power of life to reshape its environment. Cyanobacteria, tiny and ancient, were the architects of a planetary transformation that enabled the evolution of animals, plants, and ultimately humanity. Their story is a reminder of life's ability to influence the world on a planetary scale—a theme that resonates today as humans, the most recent agents of change, alter the Earth in unprecedented ways.

In studying the Oxygen Revolution, we gain insight not only into Earth's history but also into the broader dynamics of planetary habitability. The presence of oxygen is often used as a key indicator in the search for life on exoplanets, yet the Oxygen Revolution reveals that oxygen's emergence is neither inevitable nor straightforward. It is the result of a complex interplay of biology, chemistry, and geology, shaped by chance and contingency as much as by necessity.

The Oxygen Revolution transformed Earth from a barren, microbial world into a vibrant, complex biosphere. It was a turning point that set the stage for the rise of animals, the expansion of ecosystems, and the eventual emergence of intelligent life. It reminds us that Earth's atmosphere, so easily taken for granted, is the product of billions of years of evolution and interaction—a fragile, dynamic system that sustains all we hold dear.

Chapter 7: Earth in Myth and Science

Throughout human history, Earth has been more than just a home; it has been a source of mystery, inspiration, and meaning. From the earliest myths to the latest scientific discoveries, humanity's relationship with Earth has evolved in profound ways. At first, Earth was seen as a realm of divine forces, shaped by gods and spirits. Later, it became an object of inquiry, its workings decoded by observation, experimentation, and reason. Yet, even as science has illuminated its processes, Earth retains an aura of wonder—a living, breathing planet whose complexity continues to challenge our understanding.

In ancient mythologies, Earth was often personified as a deity or mother figure, reflecting its central role in sustaining life. In Greek mythology, **Gaia** was the primordial goddess of the Earth, the origin of all life and the foundation upon which the heavens and oceans rested. Similarly, in Hindu tradition, **Prithvi** is the goddess of the Earth, often depicted as nurturing and steadfast. Indigenous cultures around the world, from the Native American traditions of Turtle Island to the Australian Aboriginal concept of the Dreamtime, view Earth not as an inert object but as a living entity, interconnected with all beings.

These myths are more than stories; they are reflections of humanity's dependence on and reverence for the natural world. They highlight the ancient recognition that Earth is not merely a backdrop for life but an active participant in it, a provider and protector whose health is intertwined with our own.

As human understanding grew, so did our questions about Earth's place in the cosmos. Early civilizations believed Earth was the center of the universe, a view codified in the geocentric models of Aristotle and Ptolemy. This vision of a stationary Earth, orbited by the Sun, Moon, and stars, reflected not just scientific assumptions but also humanity's sense of importance and centrality.

The Copernican Revolution in the 16th century shattered this worldview, placing the Sun at the center of the solar system and relegating Earth to

Earth

a planet among planets. This shift was more than an astronomical revelation; it was a profound philosophical challenge, forcing humanity to reconsider its place in a vast, seemingly indifferent universe.

With the advent of telescopes, the true nature of Earth began to emerge. Galileo's observations of the Moon, planets, and stars revealed a cosmos governed by physical laws, rather than divine intervention. This new understanding extended to Earth itself, as scientists like Newton and Kepler described its motion, gravity, and place within the solar system.

In the centuries that followed, Earth became an object of rigorous scientific study. Geologists unraveled its history, discovering that the planet was billions of years old and shaped by slow, powerful forces like erosion, volcanism, and tectonic activity. Biologists traced the origins of life, linking the diversity of species to evolutionary processes shaped by Earth's changing environments. Chemists and physicists explored the composition of the atmosphere, the properties of water, and the cycles of energy that sustain ecosystems.

By the 20th century, a new vision of Earth emerged—one that emphasized its interconnected systems. The discovery of plate tectonics revealed the dynamic nature of the planet's crust, while advances in ecology highlighted the delicate balance of its biosphere. The first images of Earth from space, taken during the Apollo missions, reinforced this holistic perspective. The sight of a blue and white sphere, suspended in the blackness of space, became an icon of both beauty and vulnerability.

These images, often referred to as the "Blue Marble," transformed how humanity saw its home. They underscored the singularity of Earth in the vastness of the cosmos, inspiring movements for environmental conservation and global unity. For the first time, humanity could see the planet as a whole—a fragile oasis of life in an otherwise barren solar system.

Earth's dual identity as a scientific object and a cultural symbol continues to shape our relationship with it. Science reveals its processes: the carbon cycle that regulates its atmosphere, the tectonic forces that build its mountains, and the biological networks that sustain its ecosystems. At the same time, Earth remains a source of wonder, its beauty celebrated in art, literature, and spirituality.

Earth

This tension between scientific understanding and emotional connection is perhaps most evident in the Anthropocene, the current epoch in which humans have become a dominant force shaping the planet. Our knowledge of Earth's systems gives us unprecedented power to alter them, from the climate to the oceans to the land. Yet this power also brings responsibility—a recognition that our actions ripple through the interconnected systems that sustain life.

The myths of Earth as a nurturing mother or a sacred being remind us of our dependence on its generosity, while science reveals the mechanisms that underpin its vitality. Together, these perspectives challenge us to balance reverence with reason, wonder with stewardship.

Earth's story is far from over. As we explore other planets and search for habitable worlds beyond our solar system, Earth remains our baseline for understanding what makes a planet alive. It is the only world we know that harbors life, a living paradox where stability and change coexist, where life shapes the planet even as it is shaped by it.

In the end, Earth is more than just a place; it is a partner in the grand experiment of existence. Its myths remind us of our connection to something greater, while its science reveals the intricate dance of forces that sustain us. It is a mirror of our past, a canvas for our present, and a guide to our future—a singular world in a cosmos of possibilities.

Chapter 8: The Anthropocene Dilemma

The Anthropocene—an epoch defined by humanity's transformative impact on Earth—is a profound turning point in the planet's 4.5-billion-year history. Unlike the gradual processes of plate tectonics, atmospheric evolution, or biological adaptation, the changes wrought by human activity have unfolded with breathtaking speed. In just a few centuries, humans have reshaped the planet's surface, altered its atmosphere, and disrupted ecosystems on a global scale. The Anthropocene is both a testament to human ingenuity and a stark reminder of the fragility of the systems that sustain life.

The term "Anthropocene" was popularized by Nobel laureate Paul Crutzen in the early 2000s to describe a new epoch in which human influence rivals or exceeds natural forces in shaping the Earth. Though the term has not yet been officially adopted by the scientific community, it encapsulates the reality that no part of the planet remains untouched by humanity. From urban sprawl to deforested rainforests, from melting glaciers to acidified oceans, the fingerprints of human activity are everywhere.

The roots of the Anthropocene can be traced to the **Industrial Revolution**, when fossil fuels like coal and oil began to power machinery, transportation, and industry on an unprecedented scale. This period marked the beginning of humanity's large-scale alteration of the carbon cycle. By releasing massive amounts of carbon dioxide into the atmosphere, industrialization initiated the process of **global warming**, a phenomenon that now threatens the stability of Earth's climate.

Yet, the Anthropocene is about more than just carbon. It encompasses a host of interconnected changes, including deforestation, biodiversity loss, soil degradation, and pollution. Forests, which act as carbon sinks and support countless species, have been cleared at an alarming rate for agriculture, timber, and urban development. This deforestation not only contributes to climate change but also disrupts water cycles, accelerates soil erosion, and fragments habitats.

Earth

The loss of biodiversity is another defining feature of the Anthropocene. Scientists estimate that species are disappearing at a rate 1,000 to 10,000 times higher than the natural extinction rate, driven largely by habitat destruction, overexploitation, pollution, and climate change. This **Sixth Mass Extinction**, as it is sometimes called, threatens the integrity of ecosystems that provide essential services, from pollination to nutrient cycling to climate regulation.

Human activity has also profoundly altered the hydrological cycle. Rivers have been dammed, diverted, and polluted to supply water for agriculture, industry, and cities. Groundwater, a critical resource for millions, is being depleted faster than it can be replenished, while glaciers and snowpacks that feed rivers are disappearing due to rising temperatures. Oceans, too, bear the brunt of human impact. Overfishing has depleted stocks of key species, while warming waters and acidification threaten coral reefs, which serve as nurseries for marine life.

Perhaps the most visible marker of the Anthropocene is the accumulation of **plastic pollution**. Plastics, which did not exist a century ago, are now ubiquitous, from the deepest ocean trenches to the most remote mountaintops. Microplastics have entered the food chain, their long-term effects on ecosystems and human health still unknown. These synthetic materials are a stark reminder of humanity's ability to create durable but disruptive artifacts that outlast the systems they exploit.

At the heart of the Anthropocene lies a profound dilemma: how can humanity reconcile its need for growth, progress, and innovation with the finite limits of a planet that sustains us? The very systems that enabled human civilization—agriculture, energy production, industrialization—are the same systems driving environmental degradation. Our species, uniquely capable of altering the planet, now faces the challenge of navigating the consequences of its own success.

The Anthropocene is as much a philosophical reckoning as it is a scientific reality. It forces us to confront the duality of human nature: our capacity for ingenuity and creation, and our tendency toward short-sightedness and destruction. It raises questions about responsibility and stewardship: What do we owe to the planet that has nurtured us? What do we owe to future generations who will inherit its challenges?

One of the most critical aspects of addressing the Anthropocene dilemma is recognizing the interconnectedness of Earth's systems. Climate, biodiversity, water, soil, and human activity are not separate domains; they are deeply intertwined. A solution to one problem—such as transitioning to renewable energy—can have cascading benefits for others, such as reducing deforestation and protecting ecosystems. Conversely, failure to address one issue, such as biodiversity loss, can undermine efforts to combat climate change or ensure food security.

The Anthropocene also calls for a reimagining of humanity's role on Earth. Historically, humans have seen themselves as separate from nature, as masters of a world that exists to be exploited. The Anthropocene challenges this narrative, highlighting the consequences of treating the planet as an inexhaustible resource. It invites a new perspective, one that emphasizes cooperation with natural systems rather than domination over them.

While the challenges of the Anthropocene are immense, they are not insurmountable. Humanity has already demonstrated its capacity for change. The global effort to address the depletion of the ozone layer in the late 20th century is a powerful example. Through international cooperation and scientific innovation, the production of ozone-depleting substances was dramatically reduced, and the ozone layer is now on a path to recovery. This success shows that collective action, guided by science and sustained by political will, can reverse even global-scale environmental crises.

The Anthropocene is also a moment of opportunity. It is a call to harness the same ingenuity that created these challenges to solve them. Advances in technology, from renewable energy to sustainable agriculture to biodegradable materials, offer pathways to a more harmonious relationship with the planet. Movements for rewilding, conservation, and circular economies demonstrate the potential for transformative change at both local and global scales.

Ultimately, the Anthropocene is a story of choices. It is a reminder that the future of Earth is not predetermined but shaped by the actions we take today. Will we continue on a path of unsustainable growth, or will we redefine progress in a way that values balance, resilience, and the well-being of all life?

Earth

The Anthropocene dilemma is a mirror for humanity, reflecting both our achievements and our shortcomings. It challenges us to rise above short-term thinking, to see ourselves as part of a larger system, and to act with foresight and compassion. It is a test of our ability to learn from the past, adapt to the present, and imagine a sustainable future.

Earth's systems, though strained, remain resilient. The planet has endured asteroid impacts, mass extinctions, and ice ages, and it will endure long after humanity is gone. But the Anthropocene is not about Earth's survival—it is about ours. It is about whether we can learn to live within the boundaries of a finite planet while honoring the systems that sustain us.

In the Anthropocene, humanity is no longer a spectator of planetary change but an actor on the global stage. The question is whether we will rise to meet the responsibility that comes with this role—or whether we will allow the forces we have unleashed to shape the future for us.

Conclusion: A Fragile Miracle

Earth is nothing short of a miracle. It is a living planet, teeming with life, sustained by a delicate balance of systems that have evolved over billions of years. Its oceans breathe with the tides, its atmosphere shields and nurtures, and its crust moves like a vast puzzle, endlessly reshaping the surface we call home. Earth is a dynamic, interconnected system where every element, from the tiniest microorganism to the tallest mountain, plays a role in maintaining the whole. Yet, this miraculous balance is also extraordinarily fragile.

In the vastness of space, Earth stands alone. Despite the discovery of countless exoplanets, none has been found that matches the precise conditions necessary to support life as we know it. Our planet occupies a unique position in the solar system, bathed in sunlight but not scorched, encased in a protective atmosphere, and endowed with liquid water. These conditions have allowed life to flourish and evolve, creating the astonishing biodiversity we see today.

But Earth's history is also a story of resilience in the face of cataclysm. It has endured asteroid impacts, mass extinctions, ice ages, and volcanic upheavals. Each time, life has adapted, emerging stronger and more diverse. These cycles of destruction and renewal are part of what makes Earth unique—a planet not only alive but capable of evolving life in the face of adversity.

Despite this resilience, the balance that sustains life is remarkably sensitive. Small changes in temperature, atmospheric composition, or ocean chemistry can ripple across ecosystems, triggering shifts that affect the entire planet. The carbon cycle, for example, is finely tuned, regulating Earth's temperature by cycling carbon dioxide between the atmosphere, oceans, and land. Disruptions to this cycle—such as those caused by human activity—can have cascading effects, warming the planet, acidifying the oceans, and altering weather patterns.

The fragility of Earth's systems is mirrored in its geological and biological processes. The same tectonic forces that renew the crust and shape the continents can unleash devastating earthquakes and eruptions. The same atmosphere that protects life from harmful radiation can become a trap if greenhouse gases accumulate unchecked. Earth's dynamism, while essential to its vitality, is also a source of vulnerability.

This duality—resilience and fragility, stability and change—defines Earth as a living paradox. It is a planet capable of withstanding the forces of the cosmos, yet so finely balanced that even minor disruptions can have profound consequences. For much of Earth's history, these disruptions came from natural processes. But in the Anthropocene, humanity has become the most significant force shaping the planet.

The Anthropocene is a reminder that Earth's systems are not infinite, nor are they impervious to human actions. The rapid pace of change driven by deforestation, pollution, climate change, and biodiversity loss threatens to tip the balance that has sustained life for billions of years. Yet, the Anthropocene also offers an opportunity to reflect on humanity's role as stewards of this fragile miracle.

Stewardship begins with understanding. To protect Earth, we must first comprehend the systems that sustain it—the cycles of carbon and water, the interplay of geology and biology, the interconnectedness of ecosystems. Science has illuminated much about how these systems work, but it has also revealed their limits. As we learn more about Earth's fragility, we are called to act not out of fear but out of awe and responsibility.

Protecting Earth requires a shift in perspective. For much of human history, we have seen the planet as a resource to be exploited, its forests, oceans, and minerals treated as infinite commodities. This mindset has driven extraordinary progress, but it has also led to environmental degradation and the loss of connection with the natural world. The Anthropocene challenges us to see Earth not as a resource but as a partner—a living system that sustains us as much as we shape it.

This shift is not merely philosophical; it is practical. Efforts to mitigate climate change, restore ecosystems, and transition to sustainable energy are not just about preserving the planet—they are about ensuring humanity's future. Earth will survive the Anthropocene, just as it has survived mass extinctions and asteroid impacts. The question is whether humanity will survive alongside it.

Earth

Earth's fragility is also a source of hope. The same systems that are vulnerable to disruption are also capable of recovery. Forests regrow, coral reefs regenerate, and species rebound when given the chance. The story of the ozone layer's recovery, driven by global cooperation to phase out harmful chemicals, shows that even planetary-scale problems can be addressed when humanity comes together with shared purpose.

From space, Earth appears as a blue jewel suspended in the darkness—a vision of unity and wholeness. This perspective, made possible by technology, has transformed how we see our planet. It reminds us that Earth is not an abstraction but a singular, vibrant world, both fragile and precious. It is a reminder of our shared responsibility to protect this home we all share.

The story of Earth is still being written. It is a story of resilience and change, of challenges and opportunities, of humanity's capacity to harm and to heal. It is a reminder that the choices we make today will shape the world we leave for future generations.

In the end, Earth is more than just a planet—it is a miracle of life, a testament to the interconnectedness of the cosmos, and a call to stewardship. It is a living paradox, a fragile balance, and a source of endless wonder. Protecting it is not just a duty; it is a privilege, an expression of gratitude for the systems that sustain us and the beauty that inspires us.

Earth's fragility is its strength. Its resilience is its hope. And its future is ours to shape.

End Note: Looking Back from Afar

From the vantage of space, Earth is a luminous beacon, a singular jewel in a vast and silent cosmos. The iconic images captured by astronauts—Earthrise over the Moon's horizon, the Blue Marble set against the void—offer a perspective both humbling and profound. Seen from afar, national boundaries dissolve, the chaos of human affairs fades, and what remains is the fragile, interconnected system that sustains life.

Earth is unique, as far as we know. Despite the discovery of thousands of exoplanets, no other world has revealed itself to be alive in the same way. Planets like Venus and Mars teach us about the forces that shape lifeless worlds, while distant gas giants and icy moons remind us of the diversity of planetary systems. Yet none of these compare to the vibrant complexity of Earth, where oceans dance with tides, forests breathe in rhythm with the atmosphere, and ecosystems teem with life.

Looking back from afar deepens our appreciation for the planet we call home. It reminds us that Earth's habitability is not a given but the product of billions of years of dynamic interactions—between the Sun and the atmosphere, between tectonic plates and oceans, between life and the planet itself. Each of these elements has played a role in shaping Earth's resilience, yet each is also vulnerable to disruption.

The act of looking back also compels reflection on humanity's role in shaping Earth's present and future. From space, the signs of human activity are visible: sprawling cities, deforested landscapes, and the glow of artificial lights. These markers of progress and innovation stand alongside evidence of strain—melting glaciers, shrinking forests, and polluted waters. The Anthropocene is etched into Earth's surface, a testament to our power and the challenges it brings.

Yet looking back also reveals the planet's enduring beauty. The swirls of white clouds, the deep blue of oceans, the greens and browns of continents—all these elements speak to Earth's vitality, a living world set apart from the barren surfaces of neighboring planets. This beauty inspires not only awe but also a sense of responsibility. The privilege of witnessing Earth from space comes with the realization that it is not just a home but a rare and precious phenomenon.

Earth

As we search for life beyond Earth, the uniqueness of our planet becomes even more apparent. The criteria for habitability—liquid water, a stable climate, and protective atmospheres—are easy to define in theory but rare to find in practice. Earth is our benchmark for understanding what it means for a planet to be alive, and its lessons inform our exploration of exoplanets and distant stars.

But the search for other worlds is not just about finding another Earth. It is also about understanding what makes our planet special and why it has flourished while others have faltered. Venus offers a cautionary tale of a runaway greenhouse effect, while Mars reveals the consequences of losing an atmosphere and water. These worlds remind us that habitability is a delicate balance, one that Earth has maintained through both chance and resilience.

Looking back from afar also connects us to the broader story of the cosmos. Earth, with its life and complexity, is not an isolated phenomenon but a product of the same forces that shaped the stars, galaxies, and universe itself. The carbon in our bodies, the oxygen we breathe, and the water we drink all trace their origins to ancient stars that exploded long before Earth existed. This cosmic perspective highlights the interconnectedness of life, matter, and energy, linking us to the universe on a fundamental level.

Philosophically, looking back from afar challenges us to rethink our place in the cosmos. It humbles us, reminding us of our smallness in the face of the universe's vastness, but it also elevates us, highlighting the uniqueness of life on Earth and the power of humanity to understand and shape its world. It asks us to balance awe with action, wonder with responsibility.

The act of looking back is both a celebration and a call to stewardship. It celebrates Earth's singularity, its beauty, and its resilience, but it also calls us to protect what is irreplaceable. The challenges of the Anthropocene, from climate change to biodiversity loss, demand not only scientific solutions but also a shift in perspective—a recognition of Earth's value beyond its resources and a commitment to preserving its systems for future generations.

Earth

As we gaze outward, toward the stars, Earth remains our anchor. It is our reference point, our home, and our legacy. The lessons we learn from studying Earth guide our exploration of other worlds, just as the discoveries of other worlds deepen our understanding of Earth. This interplay between looking outward and looking back is at the heart of humanity's quest to understand its place in the universe.

In the end, Earth is a reminder of what is possible—a planet that has not only endured but thrived, a living system that reflects the creativity of the cosmos. Looking back from afar, we see not just a planet but a miracle, a fragile and enduring testament to the interconnectedness of life and the universe itself.

Earth's story is our story. Its future is our future. And its lessons, seen from near or far, are a guide to navigating the challenges and opportunities of the world we share.

www.ingramcontent.com/pod-product-compliance
Lightning Source LLC
Chambersburg PA
CBHW070944220526

45469CB00007B/2503